The goal of this book is for any student to learn math and capture the understanding and skill of math, so that it is a tool for them to use in their future.

Please Read This First

Complete all math practice: addition, subtration, multiplication. Take test 3-5 minutes, 5 times: multiplication, division and check all answers. You need to have 80% or 100% to start Algebra I work book.

Study Guide

Test

$$\begin{array}{r} 2569 \\ +\ 1933 \\ \hline \end{array} \qquad \begin{array}{r} 1327 \\ +\ 9833 \\ \hline \end{array}$$

$$\begin{array}{r} 28{,}097 \\ +\ 12{,}650 \\ \hline \end{array} \qquad \begin{array}{r} 5{,}547{,}610 \\ +\ 7{,}868{,}709 \\ \hline \end{array}$$

$$\begin{array}{r} 4 \\ 4 \\ +\ 3 \\ \hline \end{array} \qquad \begin{array}{r} 2 \\ 8 \\ +\ 2 \\ \hline \end{array} \qquad \begin{array}{r} 6 \\ 0 \\ +\ 9 \\ \hline \end{array}$$

$$\begin{array}{r} 70 \\ +\ 18 \\ \hline \end{array} \qquad \begin{array}{r} 23 \\ +\ 23 \\ \hline \end{array} \qquad \begin{array}{r} 23 \\ 11 \\ +\ 44 \\ \hline \end{array} \qquad \begin{array}{r} 36 \\ 24 \\ +\ 10 \\ \hline \end{array}$$

Answer Key

$$\begin{array}{r} {\scriptstyle 1\,1\,1\,\,\,} \\ 2569 \\ +\ 1933 \\ \hline 4502 \end{array}$$

$$\begin{array}{r} {\scriptstyle 1\,\,\,\,\,1\,\,\,} \\ 1327 \\ +\ 9833 \\ \hline 11{,}160 \end{array}$$

$$\begin{array}{r} {\scriptstyle 1\,\,\,\,\,1\,\,\,\,\,\,} \\ 28{,}097 \\ +\ 12{,}650 \\ \hline 40{,}747 \end{array}$$

$$\begin{array}{r} {\scriptstyle 1\,1\,1\,1\,\,\,\,\,\,\,\,} \\ 5{,}547{,}610 \\ +\ 7{,}868{,}709 \\ \hline 13{,}416{,}319 \end{array}$$

$$\begin{array}{r} 4 \\ 4 \\ +\ 3 \\ \hline 11 \end{array} \qquad \begin{array}{r} 2 \\ 8 \\ +\ 2 \\ \hline 12 \end{array} \qquad \begin{array}{r} 6 \\ 0 \\ +\ 9 \\ \hline 15 \end{array}$$

$$\begin{array}{r} 70 \\ +\ 18 \\ \hline 88 \end{array} \qquad \begin{array}{r} 23 \\ +\ 23 \\ \hline 46 \end{array} \qquad \begin{array}{r} 23 \\ 11 \\ +\ 44 \\ \hline 78 \end{array} \qquad \begin{array}{r} {\scriptstyle 1\,\,\,} \\ 36 \\ 24 \\ +\ 10 \\ \hline 70 \end{array}$$

Test

$$\begin{array}{r} 3,942,801 \\ -\ 1,694,786 \\ \hline \end{array}$$

$$\begin{array}{r} 676,869 \\ -\ 398,789 \\ \hline \end{array}$$

$$\begin{array}{r} 5000 \\ -\ 2468 \\ \hline \end{array}$$

$$\begin{array}{r} 987,654,311 \\ -\ 123,457,899 \\ \hline \end{array}$$

$$\begin{array}{r} 111,111,110 \\ -\ 96,320,651 \\ \hline \end{array}$$

Answer Key

8 13 7 9
3, ~~9~~ *1*~~4~~ *1*2, ~~8~~ *1*~~0~~ *1*1
\- 1, 6 9 4, 7 8 6
2, 2 4 8, 0 1 5

5 16 7
~~6~~ *1*~~7~~ *1*6, ~~8~~ *1*6 9
\- 3 9 8, 7 8 9
2 7 8, 0 8 0

4 9 9
~~5~~ *1*~~0~~ *1*~~0~~ *1*0
\- 2 4 6 8
2 5 3 2

5 14 13 12 10
9 8 7, ~~6~~ *1*~~5~~ *1*~~4~~, *1*~~3~~ *1*1 *1*1
\- 1 2 3, 4 5 7, 8 9 9
8 6 4, 1 9 6, 4 1 2

0 10 10 10 0 10 10
~~1~~ *1*~~1~~ *1*~~1~~, *1*~~1~~ *1*1 ~~1~~, *1*~~1~~ *1*~~1~~ *1*0
\- 9 6, 3 2 0, 6 5 1
1 4, 7 9 0, 4 5 9

Test

1-9 Times Table's Test

X	1	2	3	4	5	6	7	8	9
1									
2									
3									
4									
5									
6									
7									
8									
9									

Complete in 3-5 minutes each day for 10 days.

Answer Key

1-9 Times Tables Test

X	1	2	3	4	5	6	7	8	9
1	1	2	3	4	5	6	7	8	9
2	2	4	6	8	10	12	14	16	18
3	3	6	9	12	15	18	21	24	27
4	4	8	12	16	20	24	28	32	36
5	5	10	15	20	25	30	35	40	45
6	6	12	18	24	30	36	42	48	54
7	7	14	21	28	35	42	49	56	63
8	8	16	24	32	40	48	56	64	72
9	9	18	27	36	45	54	63	72	81

Complete in 3-5 minutes each day for 10 days.

Test

$$\begin{array}{r} 32 \\ \times\ 4 \\ \hline \end{array} \qquad \begin{array}{r} 90 \\ \times\ 8 \\ \hline \end{array} \qquad \begin{array}{r} 74 \\ \times\ 2 \\ \hline \end{array}$$

$$\begin{array}{r} 407 \\ \times\ 6 \\ \hline \end{array} \qquad \begin{array}{r} 568 \\ \times\ 2 \\ \hline \end{array} \qquad \begin{array}{r} 807 \\ \times\ 3 \\ \hline \end{array}$$

$$\begin{array}{r} 14 \\ \times 12 \\ \hline \end{array} \qquad \begin{array}{r} 27 \\ \times 16 \\ \hline \end{array} \qquad \begin{array}{r} 22 \\ \times 22 \\ \hline \end{array} \qquad \begin{array}{r} 98 \\ \times 30 \\ \hline \end{array}$$

$$\begin{array}{r} 350 \\ \times\ 61 \\ \hline \end{array} \qquad \begin{array}{r} 298 \\ \times\ 30 \\ \hline \end{array} \qquad \begin{array}{r} 307 \\ \times\ 47 \\ \hline \end{array}$$

$$\begin{array}{r} 200 \\ \times 500 \\ \hline \end{array} \qquad \begin{array}{r} 297 \\ \times 263 \\ \hline \end{array} \qquad \begin{array}{r} 420 \\ \times 383 \\ \hline \end{array}$$

Answer Key

```
  32        90        74
x  4      x  8      x  2
----      ----      ----
 128       720       148
```

```
  4        1 1        2
 407       568       807
x  6      x  2      x  3
----      ----      ----
2442      1136      2421
```

```
   14         4           22          2
  x12        27          x22         98
  ---       x16          ---        x30
   28       ---           44        ---
 +140       162         +440         00
  ---      +270          ---      +2940
  168       ---          484       ----
            432                    2940
```

```
     3             2 2            2
    350            298            4 (crossed out)
    x 61          x 30           307
  ------          ----          x 47
    350           000           1
 +21000          +8940          2149
  -----           ----        +12280
  21350           8940         -----
                               14429
```

```
                    1 1
                    3 4 (crossed out)
                    2 2 (crossed out)       1 (crossed out)
      200           297                     420
     x500          x263                    x383
   ------          ----                   ------
      000           1                   1  1260
     0000         1 2891                  33600
 +100000          17820                 +126000
  -------        +59400                  ------
  100000          78111                  160860
```

Test

$2\overline{)13}$ $4\overline{)27}$ $3\overline{)60}$

$4\overline{)72}$ $2\overline{)523}$ $6\overline{)606}$

$10\overline{)80}$ $17\overline{)84}$ $52\overline{)796}$

$26\overline{)966}$ $64\overline{)6249}$ $24\overline{)7930}$

$714\overline{)44973}$ $714\overline{)44973}$

Answer Key

```
   6 r 1
2)1 3
 - 1 2
     1
```

```
   6 r 3
4)2 7
 - 2 4
     3
```

```
  2 0
3)6 0
 - 6
  0 0
  - 0
    0
```

```
  1 8
4)7 2
 - 4
  3 2
 - 3 2
     0
```

```
  2 6 1 r 1
2)5 2 3
 - 4
  1 2
 - 1 2
    0 3
    - 2
      1
```

```
  1 0 1
6)6 0 6
 - 6
  0 0
  - 0
    0 6
    - 6
      0
```

```
      8
1 0)8 0
  - 8 0
      0
```

```
        4 r 16
1 7)78 14
     6 8
     1 6
```

```
      1 5 r 16
5 2)7 9 6
  - 5 2
    2 7 6
  - 2 6 0
      1 6
```

```
         3 5 r 6
2 6)89 19 6 6
  -   7 8
      1 8 6
  -   1 8 0
          6
```

```
              9 7 r 41
6 4)56 1112 14 9
  -   5 7 6
        4 8 9
  -     4 4 8
          4 1
```

```
      3 3 0 r 10
2 4)7 9 3 0
  - 7 2
      7 3
    - 7 2
        1 0
      - 0
        1 0
```

```
            6 2 r 705
7 1 4)4 4 9 7 3
    - 4 2 8 4
      12 11 23 13
    - 1 4 2 8
        7 0 5
```

```
          4 0 7 r 42
2 3 3)9 4 8 7 3
    - 9 3 2
        1 6 7
    -       0
        1 6 7 3
    -   1 6 3 1
            4 2
```

Table of Contents

Unknown & Variables

First Letter	Last Letter	Unique Letter
B . . . Bob	K . . . Jack	A . . . Jan
J . . . Jack	N . . . Jan	H . . . John
M. . . Martha	Y . . . Jimmy	U . . . Juan

Unknown & variables are letters that stand for a unkown or variable.

We should know what the letter stands for.

Note:

Try not use 0 (zero). Look like a O letter.
S. Look like a 5 (five) number

Equations Are Balanced

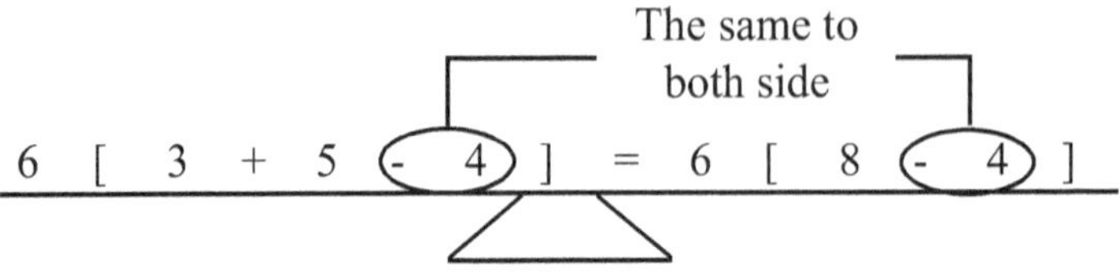

What ever you do to one side, you must do to the other to keep it balanced.

If you add, subtract, divide or multiply on one side, you must do to the other side.

Inverse Operations Opposite

Add - Subtract {opposite}

Subtract - Add {opposite}

Multiply - Divide {opposite}

Divide - Multiply {opposite}

When using inverse operations just use the opposite to complete the math problem.

Solve for Unknown Addition & Subraction

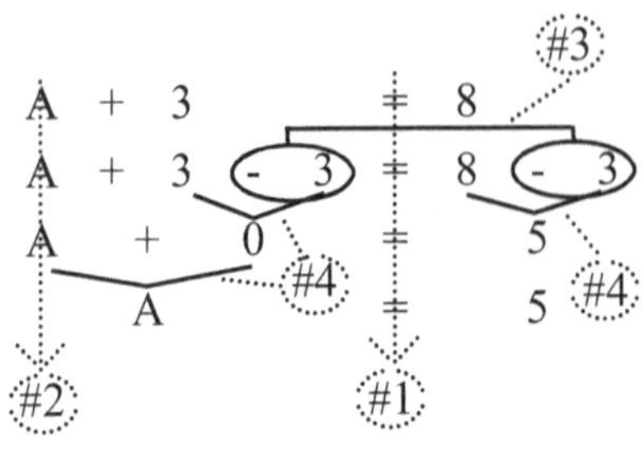

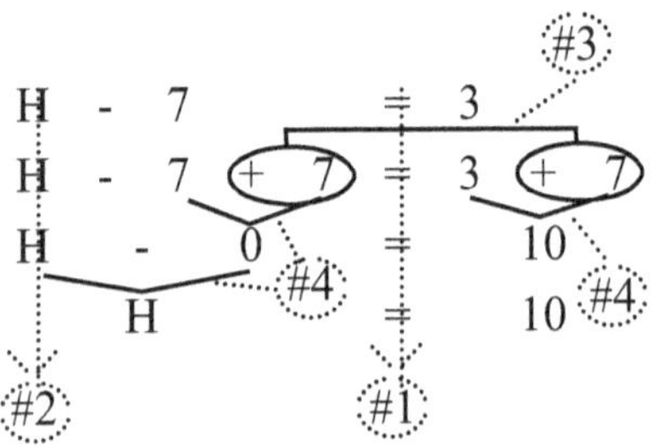

All Steps Numbered

1. Keep equal sign lined up

2. Copy directly below

3. Put inverse on both sides. Circle and connect.

4. Calculate directly below

Test #1

$A + 4 = 12$

$A - 8 = 4$

$A - 4 = 12$

$B + 3 = 24$

Answers Test #1

All work should look like this.

$$\begin{aligned} A + 4 &= 12 \\ A + 4 - 4 &= 12 - 4 \\ A + 0 &= 8 \\ A &= 8 \end{aligned}$$

$$\begin{aligned} A - 8 &= 4 \\ A - 8 + 8 &= 4 + 8 \\ A - 0 &= 12 \\ A &= 12 \end{aligned}$$

$$\begin{aligned} A - 4 &= 12 \\ A - 4 + 4 &= 12 + 4 \\ A - 0 &= 16 \\ A &= 16 \end{aligned}$$

$$\begin{aligned} B + 3 &= 24 \\ B + 3 - 3 &= 24 - 3 \\ B + 0 &= 21 \\ B &= 21 \end{aligned}$$

Solve for Unknown Multiplication & Division

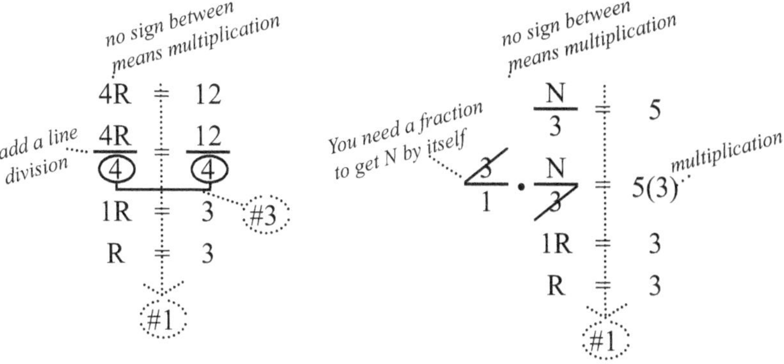

All Steps

1. Keep equal sign lined up
2. Copy directly below
3. Put inverse on both sides. Circle and connect.
4. Calculate directly below

Note: For multiplication you must put a line under math problem to divide the opposite. See 4R=12, the line under the 4R.

Test #2

$4A = 12$

$\frac{N}{4} = 12$

$6B = 18$

$\frac{A}{3} = 24$

Answers Test #2

All work should look like this.

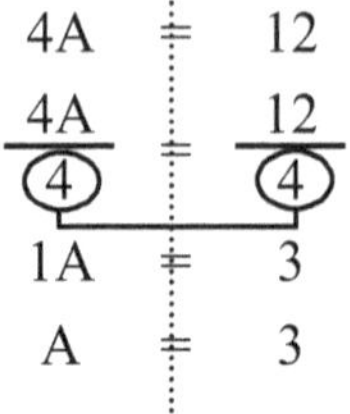

$$\frac{N}{4} = 12$$

$$\frac{4}{1} \cdot \frac{N}{4} = 12(4)$$

$$1N = 48$$

$$N = 48$$

$$6B = 18$$

$$\frac{6B}{6} = \frac{18}{6}$$

$$1B = 3$$

$$B = 3$$

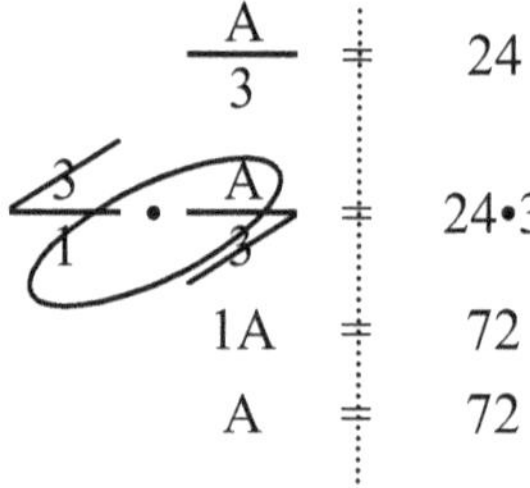

Inverse to the Other Side Only

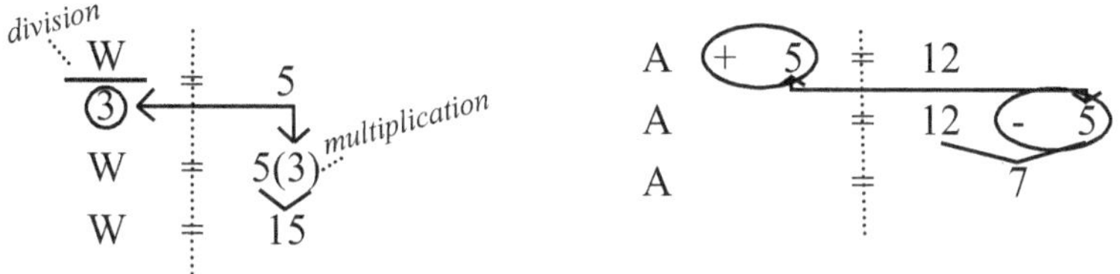

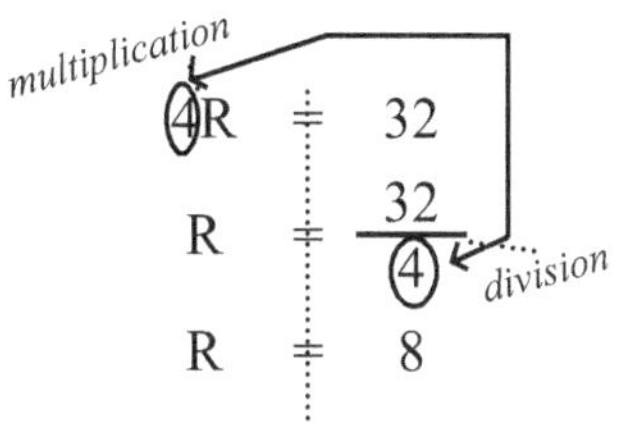

New Short Steps

1. Line up
2. Circle what I plan to move
3. Copy everything else
4. Inverse to other side. Circle & connect.
5. Calculate directly below.

Note: Look at problem. When division, use the inverse to multiplication and solve the problem. Same with addition, inverse to subtraction.

Test #3

$$\frac{W}{2} = 8$$

$$\frac{A}{9} = 3$$

$B + 8 = 25$

$A + 9 = 36$

$5R = 25$

$7R = 28$

Answers Test #3

All work should look like this.

$$\frac{W}{2} = 8$$
$$W = 8(2)$$
$$W = 16$$

$$\frac{A}{9} = 3$$
$$A = 3(9)$$
$$A = 27$$

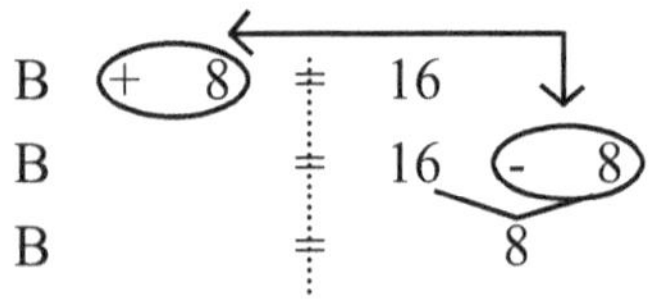

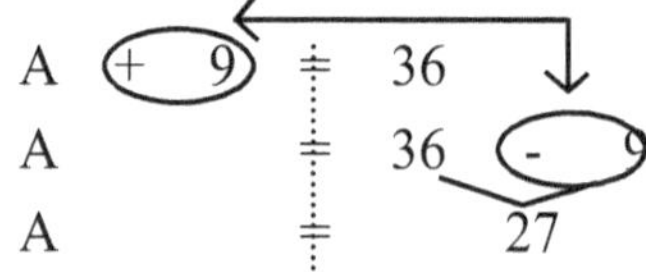

$$5R = 25$$
$$R = \frac{25}{5}$$
$$R = 5$$

$$7R = 28$$
$$R = \frac{28}{7}$$
$$R = 4$$

Combining Like Terms

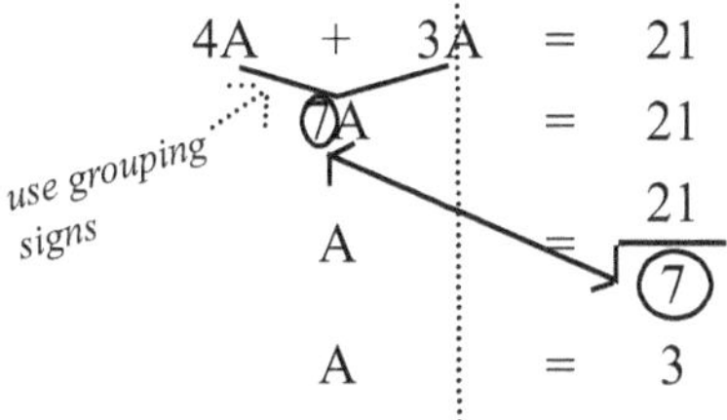

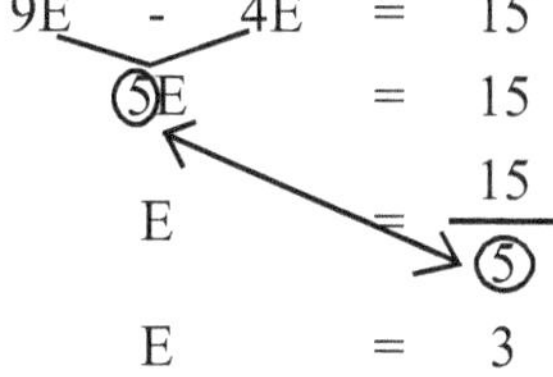

Add or Subract Like Terms

1. Use Grouping Signs for like terms.

2. Circle and connect, inverse other side.
   Add - Subtract
   Subtract - Add
   Multiply - Divide
   Divide - Multiply

3. Solve math problem

Note: Left side only has a variable only.

Test #4

$4E + 3E = 28$ $9E - 4E = 35$

$7B - 3B = 24$ $8B + 2B = 50$

Answers Test #4

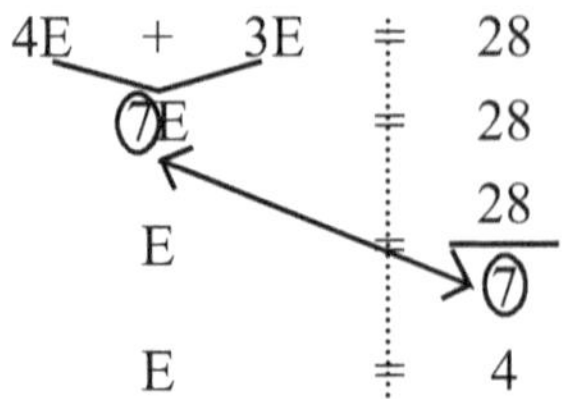

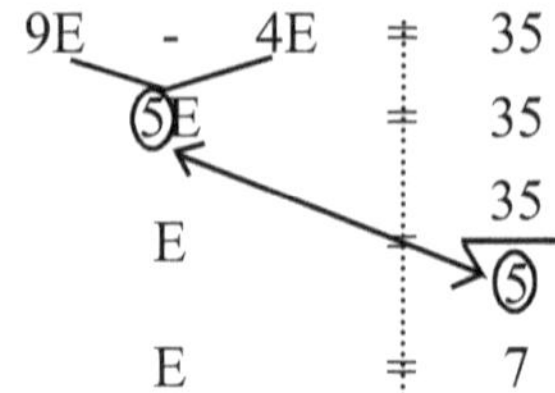

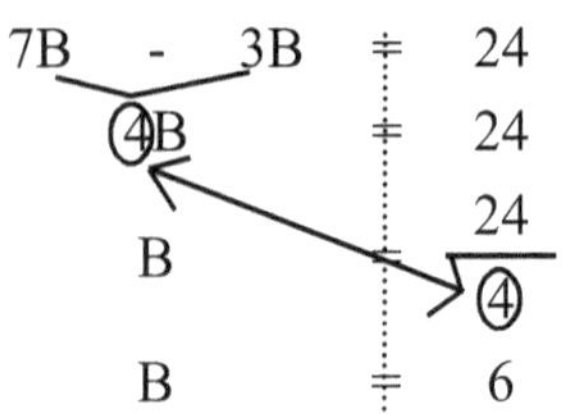

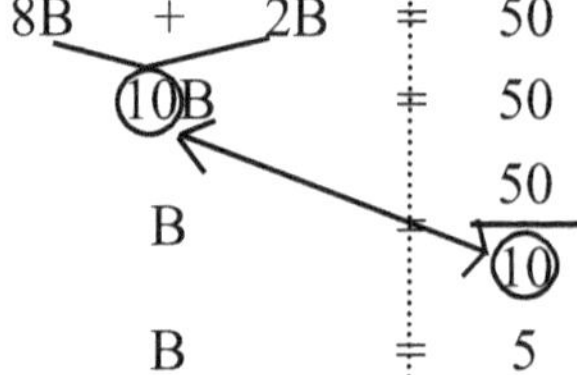

Two Inverse - Calc.

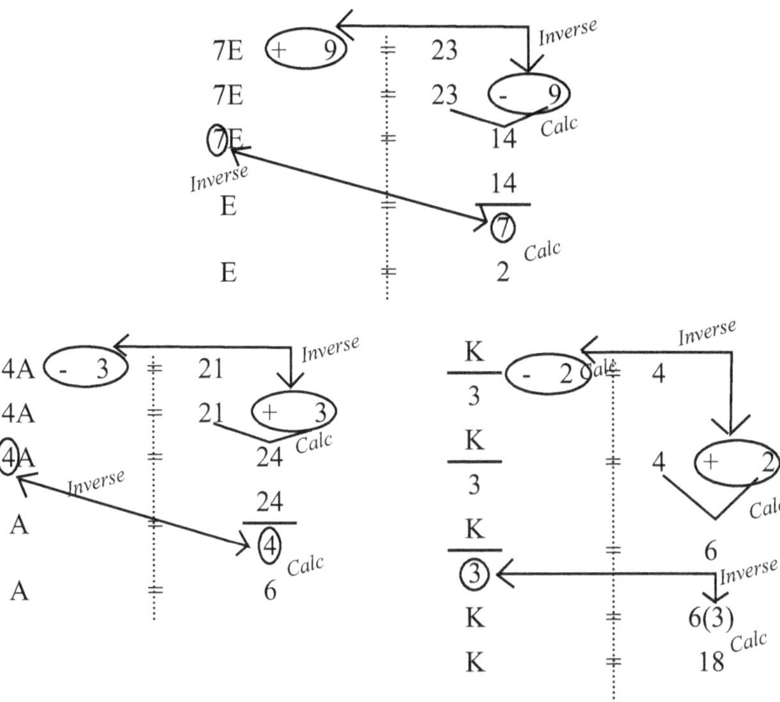

You should have 2 inverse, and 2 calc in this math problem. Inverse is the opposite

addition is subtraction
subtraction is addition
multiplication is division
division is multiplication

Backwards

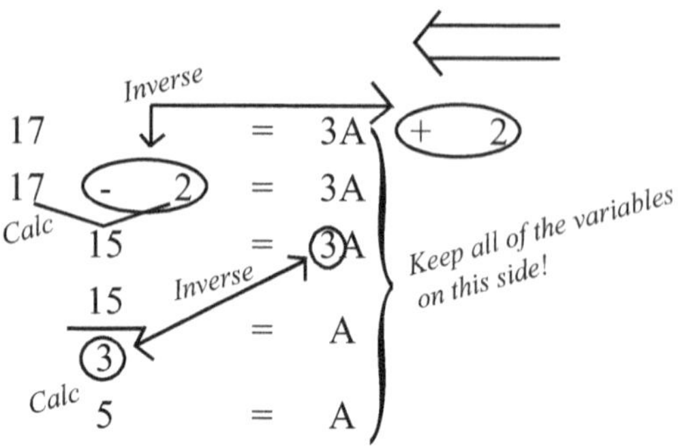

This math problem is the same as page 16, but backwards (same steps). 2 inverse; 2 cals. When going backwards you get the same answer, but you start on the right and go left. ⟸ ⟸ ⟸

Test #5

$3E + 2 = 17$

$\frac{K}{4} - 2 = 5$

$-2A - 4 = -10$

$-4A + 6 = -18$

Answers Test #5

All work should look like this.

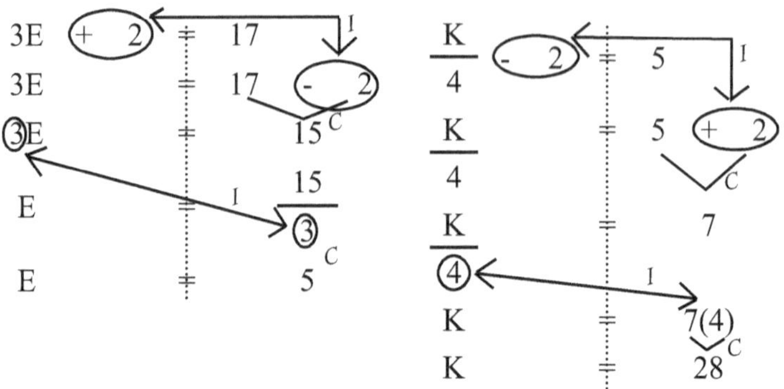

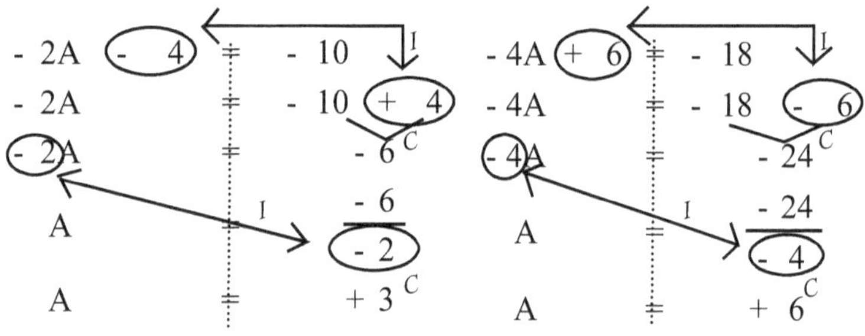

Note: Same sign +
Different sign -
Rules Intergers

I = Inverse
C = Calc.

Three Inverse - Calc.

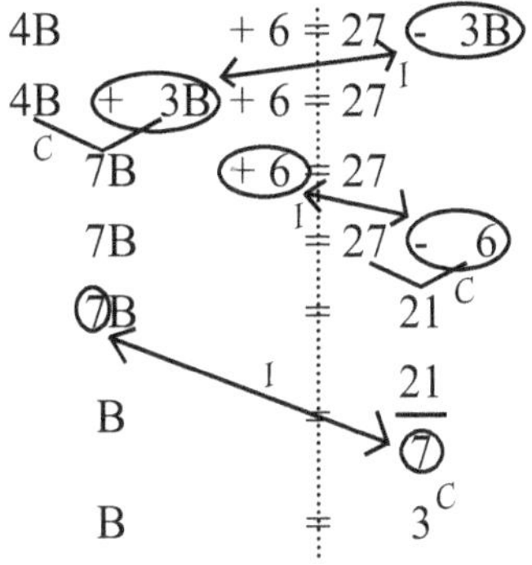

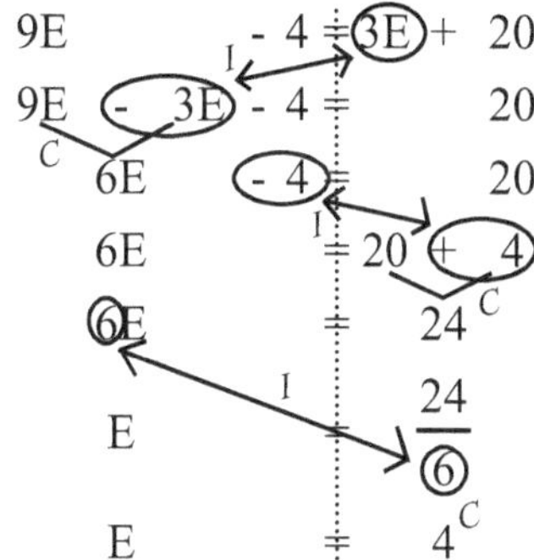

There are three steps 3 inverse & 3 calc.

Note: must line up = signs to keep neat.

I = inverse opposite (add, sub, mult, div)
C = calc. (add, sub, mult, div)

Also, keep all variables to one side.

Test #6

$8E - 5 = 5E + 10$ $\quad$ $3R - 5 = R + 7$

$5N - 9 = 5 - 2N$ $\quad$ $3W + 6 = W + 8$

Answers Test #6

All work should look like this.

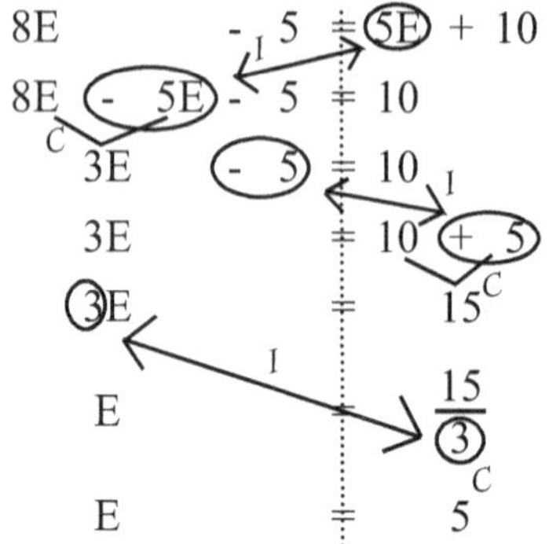

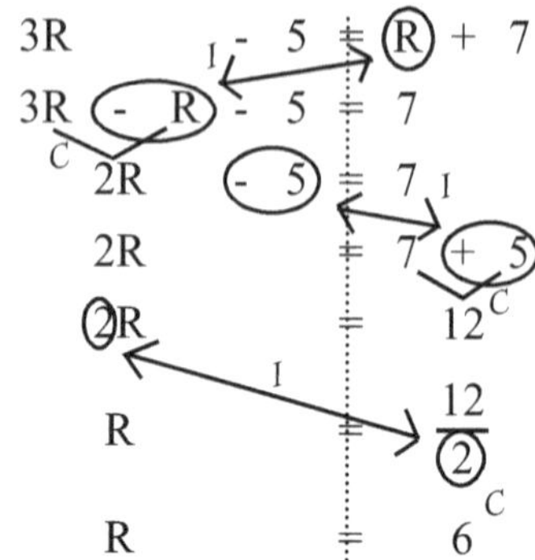

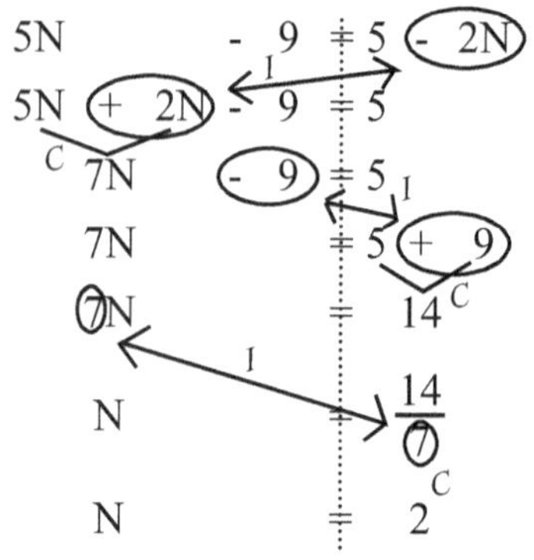

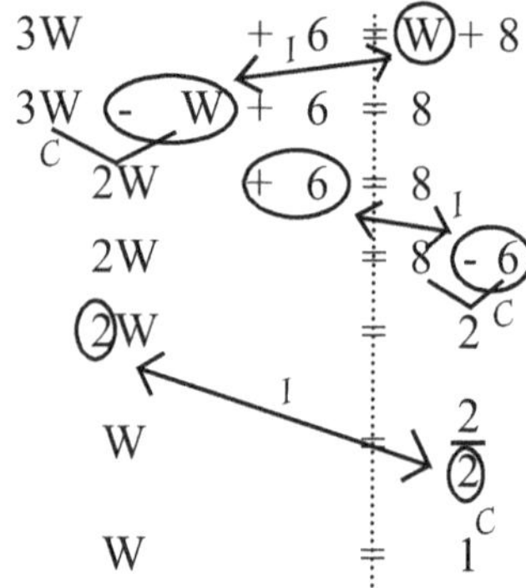

Parenthesis

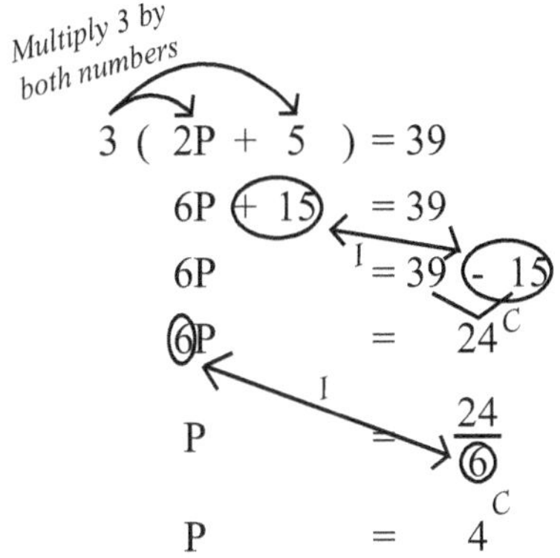

Test #7

$2(4N + 3) = 30$ $-2(4B - 9) = -6$

$3(2A + 5) = 39$ $4(3E + 5) = 80$

Answers Test #7

All work should look like this.

Rules: same sign ⊕ different sign ⊖

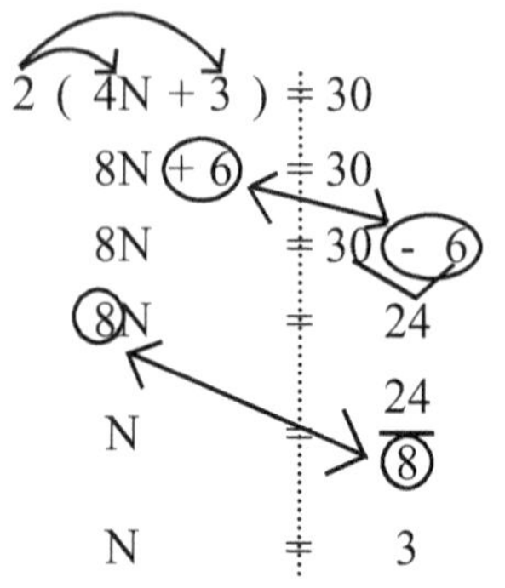

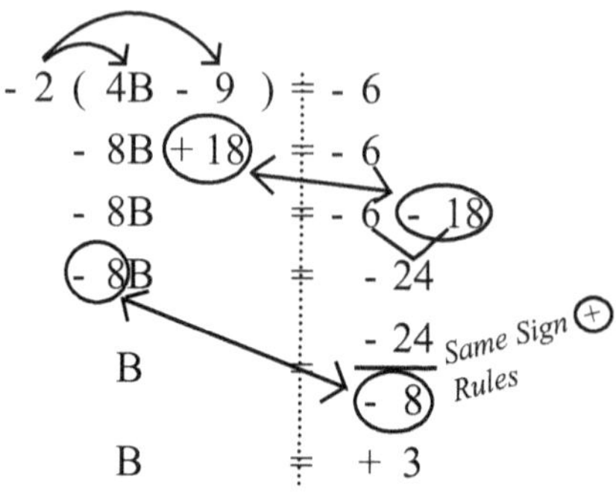

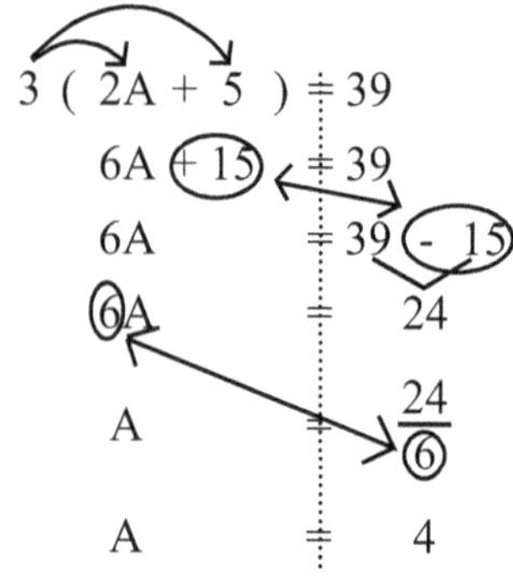

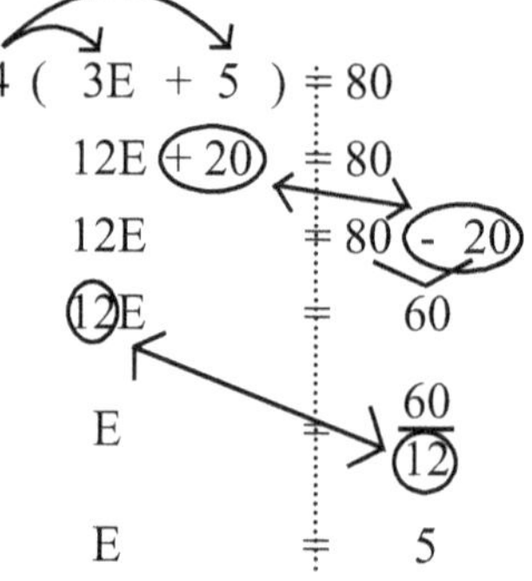

The Ugly Page

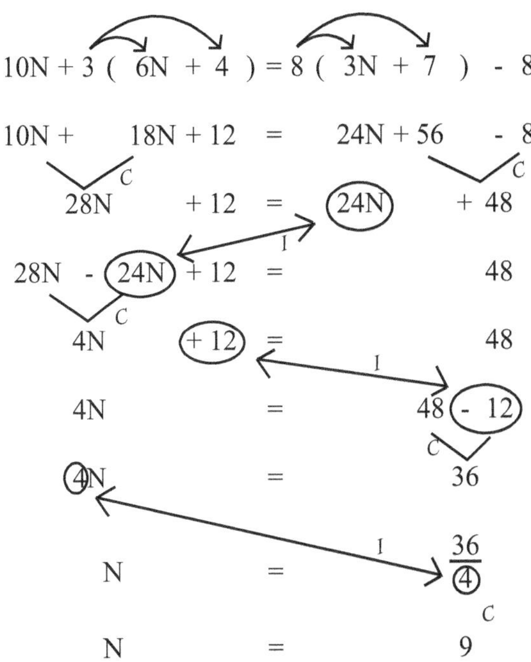

When doing this page, you will have 2 math problems with = signs between them. Combining like terms, add or sub., inverse, mult., div., inverse and calc. for answers.

Note: I = inverse
C = calc

Test #8

$$7(4B + 6) + 5 = 4(3B + 9) + 43$$

Answers Test #8

All work should look like this.

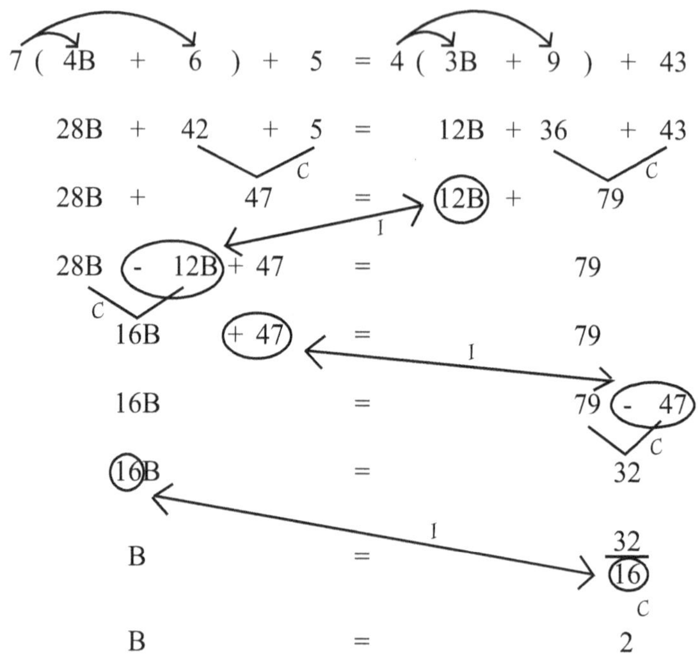

Note: I = inverse
C = calc.

Combining Signed Numbers
Add & Subracting

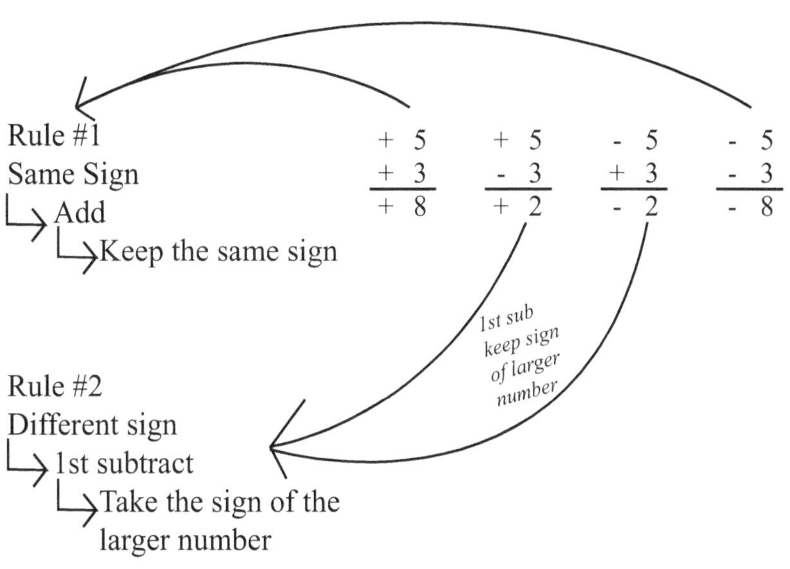

When signs are the same add up and keep the ⊕ or ⊖ sign

When the signs are different 1st sub, then keep the sign of the larger number ⊕ or ⊖ signs.

Mult. & Div. With Signed Numbers

Rule #3 Same Sign ⊕ + Keep the same direction

Rule #4 Different sign ⊖ - Do the opposite direction

Multiplication

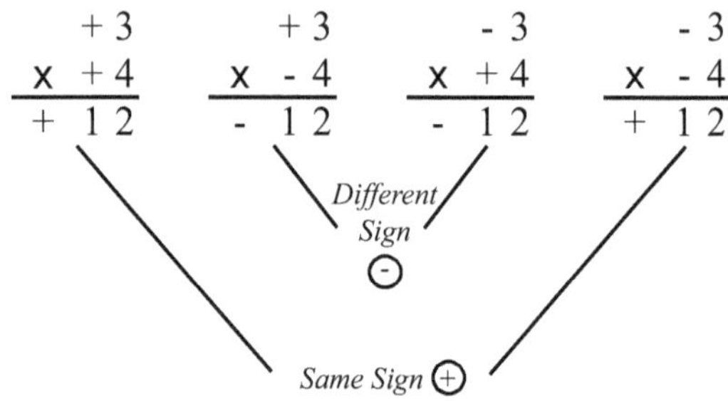

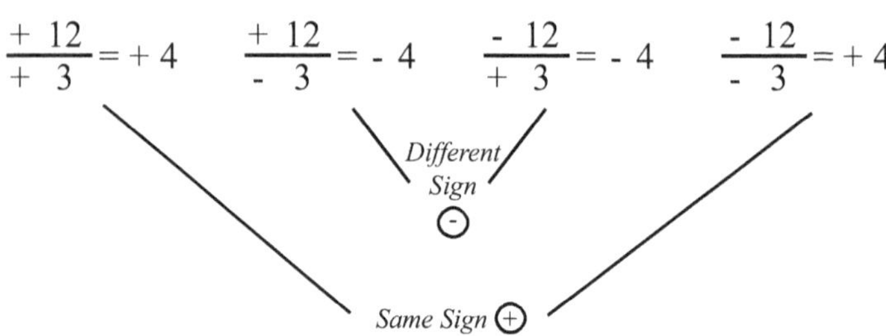

Test #9

1. (-1) (+1) (-1) (+1) (-1) (+1) (-1) = ☐

2. (-1) (-1) (-1) (-1) (-1) = ☐

3. (+1) (-1) (-1) (-1) (+1) = ☐

4. (-1) (+1) (+1) (-1) (-1) (-1) (+1) (-1) = ☐

5. (+1) (-1) (+1) (-1) (-1) (-1) (-1) = ☐

6. (-1) (-1) (+1) (+1) (-1) (-1) (+1) (+1) = ☐

Answers Test #9

1. (-1) (+1) (-1) (+1) (-1) (+1) (-1) = **+1**

   (-1) = +1 = +1 = -1 = -1

2. (-1) (-1) (-1) (-1) (-1) = **-1**

   (+1) = -1 = +1

3. (+1) (-1) (-1) (-1) (+1) = **-1**

   (-1) = +1 = -1

4. (-1) (+1) (+1) (-1) (-1) (-1) (+1) (-1) = **-1**

   (-1) = -1 = +1 = -1 = +1 = +1

5. (+1) (-1) (+1) (-1) (-1) (-1) (-1) = **-1**

   (-1) = -1 = +1 = -1 = +1

6. (-1) (-1) (+1) (+1) (-1) (-1) (+1) (+1) = **+1**

   (+1) = +1 = +1 = -1 = +1 = +1

Sign Travel for Multi & Div Only

Stay the Same

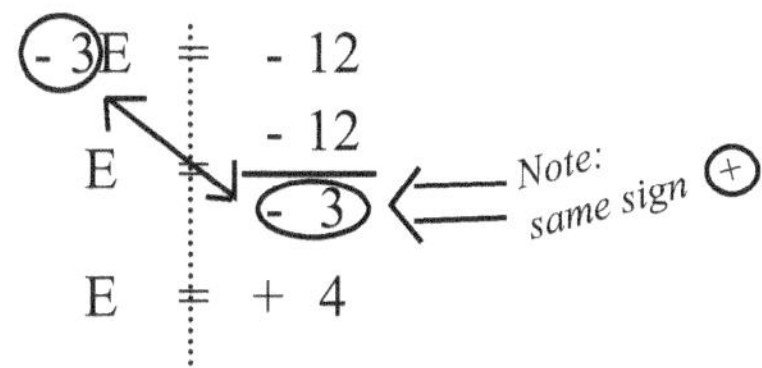

Rule 5

The sign travels for mult./div. only and stays the same.

Test #10

$-4W = -24$

$\frac{N}{-3} = 5$

$-8R + 3R = 40$

$-4K + 3 = -9$

$-5N + 18 = 3$

$\frac{N}{-6} - 3 = -5$

Answers Test #10

All work should look like this.

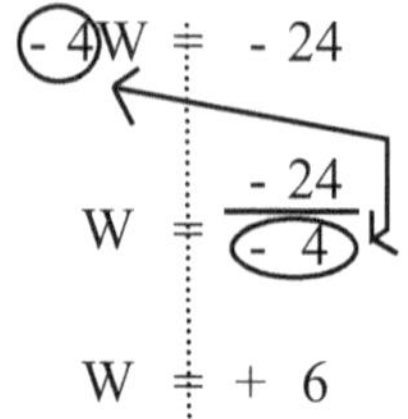

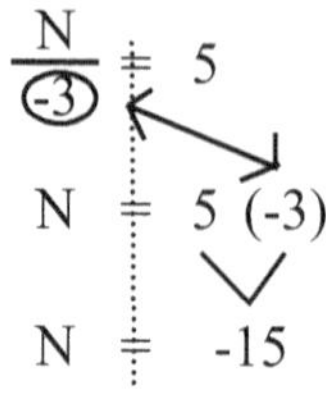

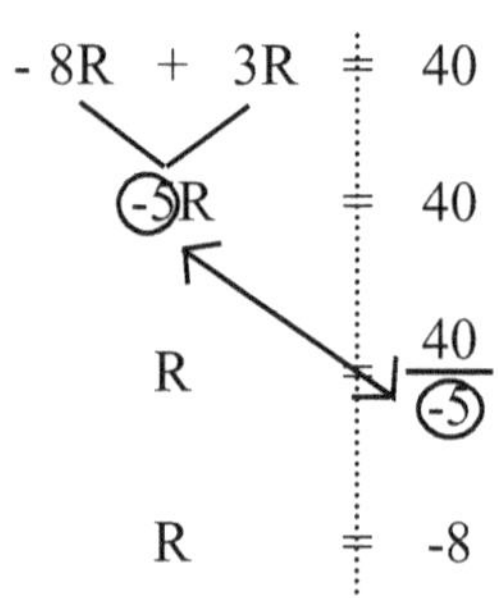

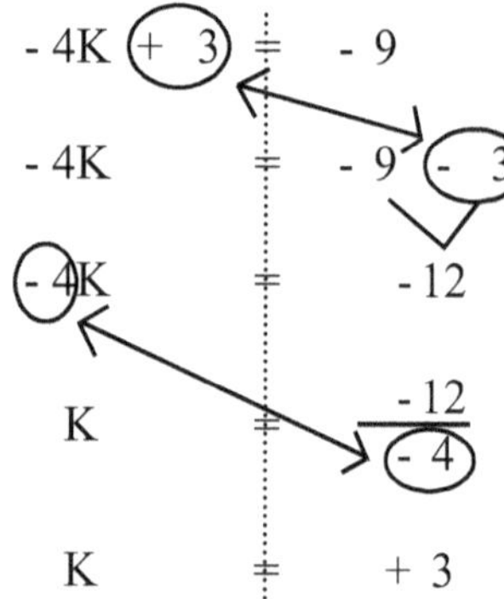

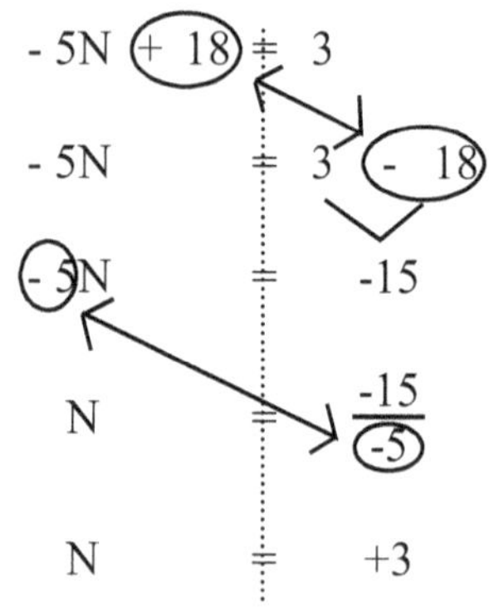

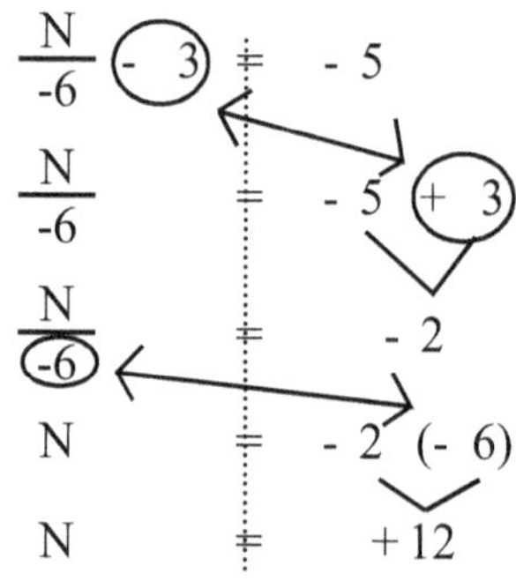

QwestHomework.com

Send order form and payment to:

11530 Lakewood Blvd.
Downey, CA 90241

voice: (562) 861-9119
fax: (562) 861-9989

I, Tony Taylor, dedicate this book to
Denise Taylor, Chantel Taylor, Tony Taylor II

www.ingramcontent.com/pod-product-compliance
Lightning Source LLC
Chambersburg PA
CBHW051247170526
45165CB00004B/1605

* 9 7 8 1 4 1 9 6 1 5 1 3 9 *